ORTHOPEDIC SURGICAL EMERGENICIES

Salman Assad, Avais Raja, Hana Manzoor, Waqar Muhammad Jan

FEBRUARY 18, 2016
HEWLETT-PACKARD

Table of Contents

General Overview of Fractures

Lower Limb

Fractures

Dislocation

Ligamentous Injury

Upper Limb

Fractures

Dislocation

Ligamentous Injury

Vertebral Column

Fractures

Dislocation

Open Fractures

Compartment Syndrome

Infections

Osteomyelitis

Septic Joint

General Overview of Fractures

What is a Fracture?

A fracture is a break in a continuity of a bone. The break may be partial or complete through the bone with or without displacement. At times the overlying soft tissue and skin may be intact known as a closed fracture or breached known as an open fracture.

Aetiology

A bone may be fractured due to a high energy trauma to a strong, healthy bone, or repetitive stress, low energy trauma on an abnormal weak bone as in the case of osteoporosis, osteomalacia, and metastasis to a bone. Direct, indirect trauma as well twisting and bending forces are responsible for a bone fracture too.

Describing a Fracture

- Name of the affected bone
- Area of the bone involved eg. Proximal, middle, distal thirds of a bone
- Direction of the fracture line eg. Transverse, Oblique, Spiral, Comminuted, Impacted
- Is the fracture simple or in fragments?
- Is the fracture stable or unstable relative to its inclination to displace after reduction?
- Associated soft tissue injury: simple vs compound; complicated vs uncomplicated
- Relation of the fracture fragments to each other in terms of alignment, rotation, translation, angulation and apposition
- Nearby joint dislocation, subluxation, and diastasis

Checklist for Fracture Description

A description of the fracture is necessary in order to pass vital information on to fellow residents and attendings. Describing a fracture may seem hard at first but by following a checklist may make it simple.

History and Physical Examination

- Musculoskeletal problems are usually positive if the patient complaints of pain, stiffness, in the muscles and joint, difficulty in performing daily activities such as walking and dressing oneself.
- The events that took place during the accident or that lead to the accident.
- Associated pain and inability to move joint(s) is determined.

The effected limb may have an obvious deformity or bruising on inspection. On palpation tenderness, swelling, erythema, limb length discrepancies and neurovascular deficits maybe assessed. Performance of various manoeuvres may be used to assess instability of the surrounding joints creating a high suspicion of associated subluxation or dislocation of a joint.

Diagnosis

Based on the history and physical examination, a suspicion of a fracture can be made but images must be obtained in order to reach a definitive diagnosis. Imaging includes plain x-rays in various views depending on the region involved, CT scan, MRI and bone scans.

Treatment

In order to manage fractures, several steps must be followed in order to achieve it.

- Reduce the fracture (open/closed).
- Fixation of the fracture (externally/internally).
- Maintain reduction and/or fixation until the fracture heals.
- Ensure the functionality of the limbs and joints involved by non-weight bearing, weight bearing exercises.

Non-operative treatment for fractures includes bed rest, pain relief, placement of a cast, placement of splints and traction.

Operative treatments include external and internal fixation. External fixation is composed of an external frame with pins and wires that are attached to the bone. It is minimally invasive and can be applied almost anywhere but unfortunately it is more prone to infections, mal-union and is problematic for the patients. External fixation maybe applied temporarily until internal fixation is performed such as in cases for open fractures and multiple traumas. Internal fixation can be done with intramedullary devices, plates, screws and wires. Intramedullary nails are applicable in the long bones of the lower limb due to the increased diameter of the bones. These nails can be inserted with minimal invasiveness and are first line treatments to restore the normal anatomical configuration of the bone. Infections and mal-union are less common is intramedullary nailing. Plating can be used for intra-articular fractures, metaphyseal fractures and long-bone fractures of the upper limb proving to be a more invasive technique. Internal fixation maybe used in cases of displaced intra-articular fractures, long –bone fractures and fractures associated with neurovascular injury.

Complications

Complications associated with surgically managing fractures include compartment syndrome, neurovascular injuries, infections (osteomyelitis), non-union fractures (failure of fracture healing), mal-union fractures, osteoarthritis, avascular necrosis, complex regional pain syndrome, deep vein thrombosis, pulmonary embolism and fat embolism.

Pelvic Fractures

Fractures of the pelvis occur due to high energy trauma from motor vehicle accidents, falls from heights and crush injuries. Fractures in this region are responsible for extensive haemorrhage, genitourinary damage and neurovascular injury which may or may not lead to sepsis and/or ARDS (Adult respiratory distress syndrome)

Patient complains of pain in the groin region, climbing stairs, putting on shoes and sexual intercourse along with difficulty in performing other daily activities which also result in pain. Bruising or swelling should be looked for in the lower abdominal and perineal area, however general condition of the patient and severe blood loss should be given first priority.

- Red flags: Intraperitoneal bleeding is confirmed when abdomen is palpated leading to irritation. Urethral damage if prostate is abnormally high on DRE. Inability to void or blood in urine suggests ruptured bladder and urethral rupture.
- Radiographical images include: AP view of the pelvis along with chest x-ray. Additional views include AP, inlet view, outlet view and right and left oblique views. CT scan is indicated when severe injury is suspected. IV urogram and/or urethrogram undertaken in severe lower abdominal injury and urethral rupture.

Types of Pelvic Fractures
1. Isolated fractures with intact pelvis ring: (Avulsion, Direct and Stress fractures)
2. Pelvic ring fractures (stable or unstable)
3. Fractures of the acetabulum (ring fractures): (Column, Transverse and Complex fractures)
4. Sacrococcygeal fractures

1. Isolated Fractures

Type	Mechanism of Injury	Diagnosis/Treatment
Avulsion Fractures (seen in athletes/sportsmen)	Violent muscle contraction leading to pulling of small segment of bone (Sartorius pulls off ASIS, Rectus femoris pulls off AILS, Adductor longus pulls off pubis and Hamstrings part of ischium).	Diagnosis of tumor (if present) on biopsy of callus. Treatment – bed rest. Surgical intervention (open reduction, internal fixation only indication on persistent symptoms of avulsion of ischial apophysis.
Direct Fractures (fall from height)	Direct blow to pelvis. Fracture of ischium/iliac blade.	Bed rest
Stress Fractures (osteoporotic/osteomalacics, long distance runners)	Pubic rami fractures. Fractures around SI joint.	Radioisotope scans for obscure stress fractures. Bed rest.

2. Pelvic Ring Fractures

Due to anatomical rigidity of the pelvis, one break point may lead to a subsequent break in the pelvis which is usually not seen because of immediate reduction or not seen in the case of children in ring fractures where SI joints are springy.

Mechanism of Pelvic ring injury (**Young & Burgess Classification**)	Description
• Anteroposterior Compression Classified according to severity: -*APC-I Injuries:* Pelvic ring stable, <2cm slight diastasis and some strain on anterior sacroiliac ligaments. -*APC-II Injuries:* Pelvic ring stable. Diastasis more marked. Tear of anterior sacroiliac ligaments. -*APC-III Injuries:* Pelvic ring unstable. Tearing of anterior and posterior sacroiliac ligaments.	-Seen in pedestrian collisions. - Patient has pain on attempting to walk. -Resulting in an 'open-book injury' where disruption of the symphysis and external rotation of pubic rami occur which may be accompanied with tearing of the anterior SI ligaments.
• Lateral Compression -*LC-I injury:* Pelvic ring stable. Hallmark is transverse fracture of pubic ramus/rami. -*LC-II Injury:* Pelvic ring stable. Anterior fracture plus Iliac wing fracture on impacted side. -*LC-III Injury:* lateral compression of one iliac wing depending on side of impact with opening anteroposterior force on opposite ileum.	-Usually occurs in falls from height or side-impact vehicular collisions. -Patient has pain on attempting to walk. -Anteriorly: Fracture of pubic rami on one or both sides -Posteriorly: Fracture of ilium or sacrum. Pelvis is unstable if sacroiliac injury is present.
• Vertical Shear Pelvic ring unstable resulting in totally disconnected hemi-pelvis (as seen in *APC-III injury*)	-Occurs in falls from height onto one leg. -Patient is in a state of shock and severe pain. Unable to pass urine or blood is present in urine. Anaesthesia of one leg indicates sciatic nerve injury. -This is a severe, unstable injury where there is presence of fracture and disruption of the pubic rami and sacroiliac region. May lead to retroperitoneal haemorrhage.
• Combination Injuries	Various combinations from all the above mentioned depending on the severity of injury.

Management

Early management: Solely concentrates on general patient care. Airway stabilization along with control of severe bleeding if present should be done. Splinting is done of painful fractures. Urethral meatus and the lower limbs are also assessed for injury. After general stabilization of patient specific management is carried out.

Specific Management

Management of Severe Bleeding	-General management of haemorrhage -External fixator applied to reduce haemorrhage in unstable fracture of pelvis. -Unstable APC injuries require a pelvic binder for reduction of internal pelvic volume and closure of the 'open book' fracture. -Peritoneal aspiration/lavage done in suspicious abdominal signs followed by laparotomy, angiography and/or pelvic packing.
Management of Urethra and Bladder	-For incomplete tear a suprapubic catheter is inserted. -For complete urethral tear, suprapubic cystostomy, evacuation of pelvic hematoma and draining of bladder.
Treatment of the Fracture	-Early external fixation (for reducing haemorrhage and counteracting shock) -*Isolated fractures:* Bed rest and lower limb traction. -*Open-book injuries*: Bed rest, posterior sling or a pelvic binder to 'close the book' -*APC-III and VS injuries*: Skeletal traction with an external fixator along with bed rest for 10 weeks. Two techniques used 1. Anterior external fixation and posterior stabilization using screws along the SI joint or 2. Plating anteriorly and iliosacral screw fixation posteriorly. -*Open pelvic fractures*: External fixation

3. Fractures of the Acetabulum

These fractures are complex due to pelvic fractures occurring hand in hand with joint disruption.

The mechanism of injury is driving of the head of the femur into the pelvis usually due to falls or high energy motor vehicle accidents. Patient may present in a shocked state with severe pain on walking. Careful neurological examination and rectal examination should be assessed along with 4 x-ray views (standard ap view, pelvic inlet view and two 45-degrees oblique views.)

Tile's Universal Classification	Description
Acetabular wall Fractures	Hip instability due to anterior or posterior involvement of acetabular rim. Reduction and fixation is advised.
Column Fractures	-Anterior column fractures are uncommon. -Posterior column fractures are associated with posterior dislocation of the hip along with sciatic nerve involvement. Internal fixation is advised to obtain a stable joint.
Transverse Fractures	-Involvement of both the anterior and posterior column along with a vertical split in the obturator foramen (T-fracture). Difficult to reduce and hold reduce.
Complex Fractures	-Roof or the walls of the acetabulum are fractured along with anterior or posterior column involvement, or both. Injury is severe and there is disruption of the joint surface. It is a variant of T-fracture except the transverse part of the 'T' lies above the acetabulum. -Operative reduction and internal fixation is advised.

Treatment

- Emergency Treatment: Manage shock and reduce dislocation. Skeletal traction is applied to the distal femur with 10kg. For central hip dislocations, lateral tractions through the greater trochanter is added.
- Non-operative Treatment: Conservative treatment in the following indications: 1) Acetabular fractures with minimal displacement, 2) Distal anterior column and distal transverse fractures, 3) Both-column fracture with loss of congruity of ball and socket of the hip, 4) Elderly patient fractures and 5) Patients with medical contraindications. Hip movement and exercises are advised.
- Operative Treatment: Indication where congruity of ball and socket of the hip are not intact in unstable fractures. Open reduction is advised via the 'Kocher-Langenbach' exposure which allows excellent exposure to the posterior columns whereas the 'Anterior Ilioinguinal approach' is advisable for anterior column fractures. Both the exposures mentioned above are used in both-column and T-type fractures. These procedures are preceded by prophylactic antibiotics followed by early hip movements after surgery.

4. Sacrococcygeal Fractures

These fractures are more common in woman and are usually seen in fall on the 'butt' or a blow from to the back. Along with pain on sitting and tenderness, the patient may also experience loss of sensation over the distribution of the sacral nerves. Radiological images include an x-ray which may show us a fractured coccyx, transverse fracture of the sacrum or a sprained sacrococcygeal joint. Reduction is required on dislocation followed by early resumption of normal activity. In severe pain, excision of the coccyx is considered.

Injuries of the Hip

Dislocation of the hip is ideally called 'fracture-dislocation' due to the fact that a dislocation of this region is always accompanied by a fracture in major fragments. This is divided into posterior and anterior dislocation.

<u>Posterior Dislocation</u>

These dislocations are most commonly seen in high energy motor vehicular collisions causing the femoral to be forced out of its socket upward thrusting of the femur. A dislocation is positive when a piece of bone from the posterior aspect of the acetabulum is sheared off. Clinical presentation of the patient is leg shortening and adduction of the leg. Further radiological imaging and thorough neurovascular examination should be carried out in order to assess the severity.

Types	Thompson and Epstein classification of hip dislocations	Treatment
I	Dislocation with no more than minor chip fractures	-Reduction is stable in type 1 injuries by application of traction and maintenance for a few days. -Terminal ranges of hip movements should be avoided to prevent collapse of femoral head.
II	Dislocation with single large fragment of posterior acetabular wall	-Immediate open reduction and anatomical fixation of the detached fragment. Hip is reduced closed if surgical expertise is not available along with traction.
III	Dislocation with comminuted fragments of posterior acetabular wall	-Treated closed with removal of retained fragments on open procedure. Segment of iliac crest may be used for reconstruction if need be.
IV	Dislocation with fracture through acetabular floor	-Initially by closed reduction -Surgical intervention is indicated in instability, retained fragments or joint incongruity.
V	Dislocation with fracture through acetabular floor and femoral head	-Initially by closed reduction -Operation is indicated if femoral head fragment is

| | | unreduced, using countersunk screw for fixation of femoral head fragment. |

If reduction is delayed for more than 12 hours, then **avascular necrosis of the femoral head** should be kept in mind because it is the most common late complication reported in traumatic hip dislocations.

Classification of Avascular Necrosis according to the Ficat grading system

Type	Description
I	No radio graphic signs of AVN
II	Changes of the femoral head subchondral bone without collapse
III	Subchondral fracture with collapse
IV	Collapse of the femoral head with changes on the acetabular side.

Treatment includes surgical intervention or bone graft or vascularized fibular grafting may be done. Arthroplasty is a definitive and successful surgical intervention but usually precede to total hip arthroplasty although research on the success of vascularized fibular grafting have also resulted in positive results but research is still being done.

Anterior Dislocation

Although this dislocation is rare as compared to posterior dislocation, it may occur in individuals involved in a road or air accident causing the neck to impinge on the acetabular rim and lever of the femoral head in front of its socket. Clinical presentation of the patient doesn't show leg shortening due to the attachment of the rectus femoris. The leg is externally rotated, slight flexed and abducted. On examination the head is not missed as it is easy to feel due to the anterior bulge of the dislocated head which is moved anteriorly and superiorly.

Radiological imaging should be carried out along with assessment of the neurovascular examination. Treatment for this type of dislocation includes same as posterior dislocation except the hip should be kept adducted while being pulled and flexed upwards. Reduction is carried out and is heard and felt.

Femoral Neck Fractures

The number of femoral neck fractures have been increasing by the year, most commonly in the elderly due to increase survival and longevity of patients owing to modern medicine. Many elderly patients are affected by osteoporosis that make them vulnerable to fractures even on the slightest traumatic event whereas patients less than 50 years of age present with high energy trauma induced femoral neck fractures. Other risk factors include long-term steroid use, osteomalacia, alcoholism, diabetes mellitus, stroke, sickle cell disease, secondary metastasis from breast, prostate and lung cancers.

Femoral neck fractures are classified according to their relationship to the capsule holding the femoro-acetabular joint (hip joint) in place: 1) intracapsular and 2) extracapsular.

The clinical presentation of these fractures are according to whether the fracture is displaced or non-displaced. Displaced fractures present with inability to walk, severe pain in the hip region, externally rotated and shortened extremity. Non-displaced fractures present with pain in the hip region for a couple of days to weeks with the ability to walk.

The fracture can be identified by performing radiographically imaging of the hip in AP, Lateral and internal rotation views. An MRI or a bone scan should be considered if no fractures are detected in the radio graphical images mentioned above.

The Gardens Classification is used to classify intracapsular femoral neck fractures with displacement before reduction. (Table 1)

The Pauwel's Classification classifies the increasing instability according to increase in fracture angle, between the fracture and the horizontal in the AP view, helping to predict the outcome with a higher angle depicting a worse prognosis. (Table 2)

The most common trochanteric fractures are the intertrochanteric fractures that are located between the greater and lower trochanteric fractures. These fractures are classified according to Kyle's description of the fracture pattern. Due to the significant blood supply and it being an extracapsular fracture it rarely leads to osteonecrosis. Extracapsular fractures on the other hand are dealt surgically with indirect reduction and a dynamic hip screw (DHS).

Gardens Classification for Intra-capsular fractures of the femoral neck

Type	Description	Treatment
I	Incomplete Fracture/Valgus Impacted Femoral Head	Nonoperative: Limited weight bearing Operative: Internal fixation with multiple screws
II	Complete Fracture/Undisplaced	Operative: Internal fixation regardless of age
III	Complete Fracture with partial displacement (trabecular bone pattern of the femoral head does not line up with the acetabulum)	Operative: Young patients with good bone quality may opt for open reduction and internal fixation. Elderly patients with osteoporosis are good candidates for unipolar hemi-arthroplasty or total hip arthroplasty considering a high suspicion of avascular necrosis of the femoral head due to compromise in the blood supply (femoral neck intramedullary vessels, ascending cervical branches of medial and lateral circumflex anastomoses, ligamentum teres vessels)
IV	Compete Fracture with full displacement (trabecular bone pattern of the head does line up with the acetabulum)	

Pauwel's Classification for describing unstable femoral neck fractures

Type	Angle
I	< 30
II	30-50
III	50-70

Kyle Classification for Intertrochanteric Fractures

Type	Description
I	Undisplaced/Uncomminuted
II	Displaced, Minimal comminution, Lesser trochanteric fracture, Varus Deformity
III	Displaced, Greater trochanter fracture, Comminuted, Varus Deformity
IV	Severely comminuted, extending to sub trochanteric region

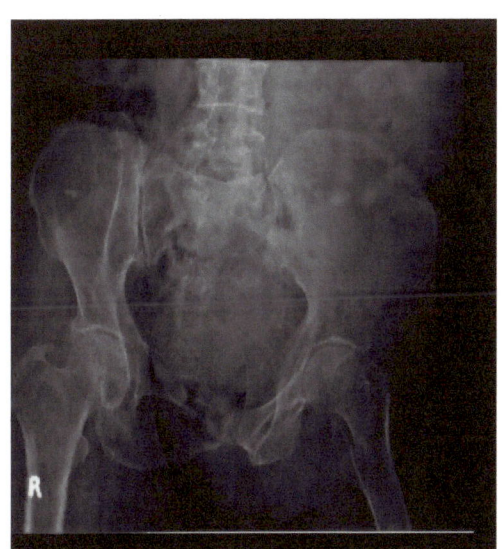
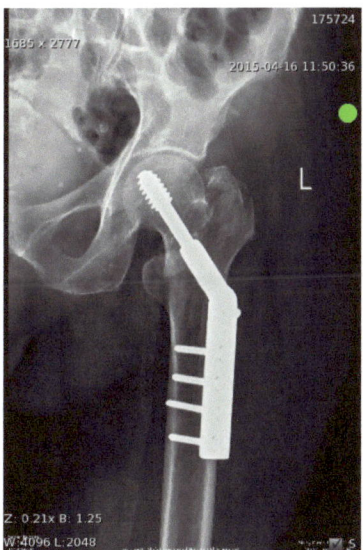
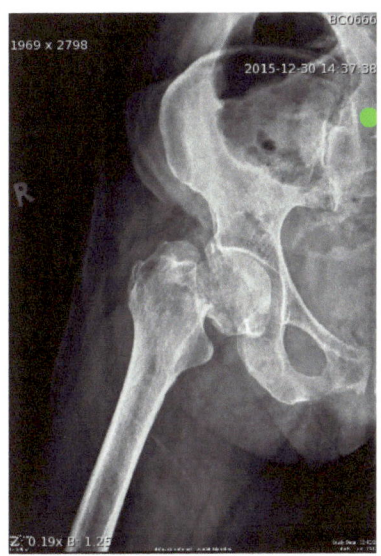
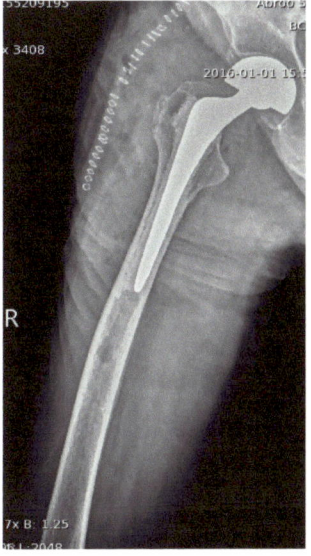

Subtrochanteric Fractures

The subtrochanteric region of the femur is relatively stronger than the femoral neck due to the presence of a wider cortex. These fractures are characterized by greater blood loss from the anastomosis between the medial and lateral circumflex vessels, extension into the intertrochanteric region, abduction, flexion and external rotation of the proximal part.

The patient has the same presentation as femoral neck fractures. AP and lateral x-rays will show a fracture below or through the lesser trochanter with the proximal segment flexed and short and the distal segment (shaft) adducted and displaced proximally.

Immediate traction should be performed to decrease the pain and blood loss followed by the definite treatment of open reduction and internal fixation with an intramedullary nail with a proximal interlocking screw. An intramedullary nail is superior to a hip screw and plate device.

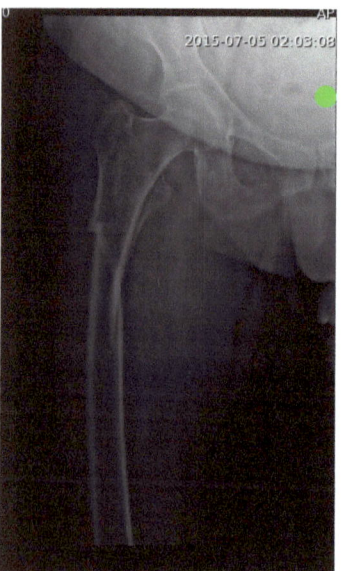
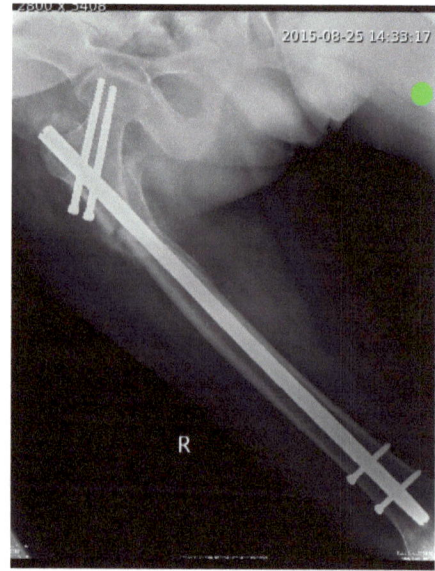

Femoral Shaft Fractures

The femoral shaft is the region 5 cm distal to the lesser trochanter and 5 cm proximal to the adductor tubercle. Fractures in this region are most commonly associated in young men after high energy impact trauma. Due to the close proximity to the profunda femoral artery, that supplies the femoral shaft, the thigh has the capacity to hold upto 4 litres of blood if the aforementioned vessel is damaged. A combination of an ipsilateral femoral neck fracture or tibial shaft fracture producing a 'floating knee' may be present.

Radiographical images include AP and lateral views of the femur as well the ipsilateral hip and knee joints. The Winquist and Hansen Classification is used to identify the amount of comminution of the fracture. A pathognomonic feature is a single large fragment known as a 'butterfly' fragment. (Table 3) In shaft fractures the proximal segment is flexed, abducted and externally rotated with the distal segment adducted.

In the emergent situation, skeletal traction with a Thomas splint must be performed within 24 hours of injury to provide pain relief, to prevent maximum blood loss and soft tissue injury and reduce the fracture temporarily until definite management is undertaken. Definitive management includes intramedullary nailing of the femur either in an antegrade or retrograde approach, nail plate and screws and external fixation.

Winquist and Hansen Classification according to the degree comminution of femoral shaft fractures

Type	Description
I	Absent or minimal comminution
II	Cortices of both fragments at least 50% contact
III	50-100% cortical comminution
IV	Circumferential comminution

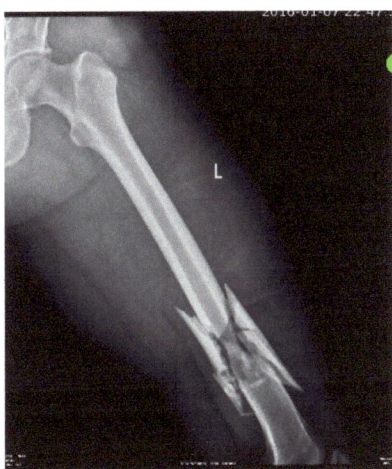
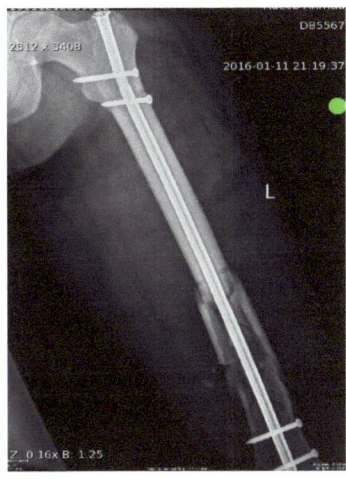

Distal Femur Fracture

Fracture of the distal femur occur in the elderly as well the young after a high energy trauma in motor vehicle accidents. Distal femoral fractures are divided into supracondylar and condylar fractures.

The location of the supracondylar region is at the junction of femoral condyles and the metaphysis of the femoral shaft. The proximal part of the fracture is pulled with the help of the quadriceps and hamstring muscles, superiorly, whereas the gastrocnemius is responsible for the posterior displacement of the distal fragment. The distal fragment may compress the neurovascular bundle in the region of the popliteal fossa resulting in diminished pedal pulses. The knee is painful and swollen due to hemarthrosis.

Radiographical studies include AP, lateral, oblique views of the distal femur along with the entire length of the femur.

Nonoperative management includes partial weight bearing of the immobilized extremity in a knee brace. Typical choice of operative management is nail plate and screws, or locking peri-articular plate being most commonly used as of now.

AO Classification of Distal Femoral Fractures

Type	Description	Surgical Management
A	Supracondylar without articular splits	Locked Intramedullary Nail or Side Plate and Screws
B	Shear fracture of one condyle	Simple Lag Screws
C	Supracondylar and Intercondylar fissures	Locked Intramedullary Nail or Side Plate and Screws

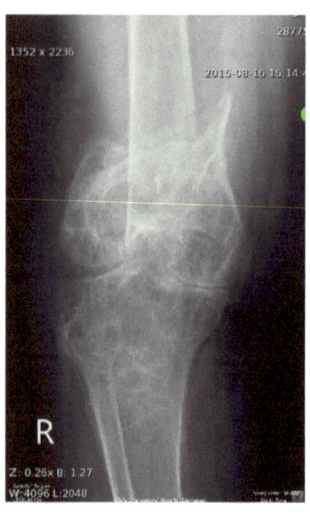
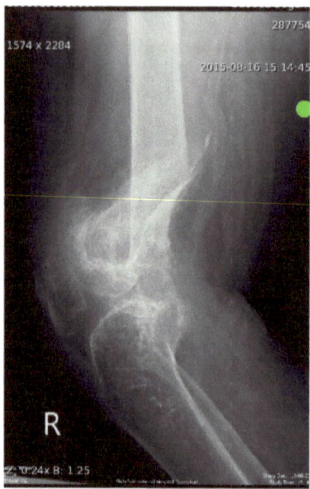
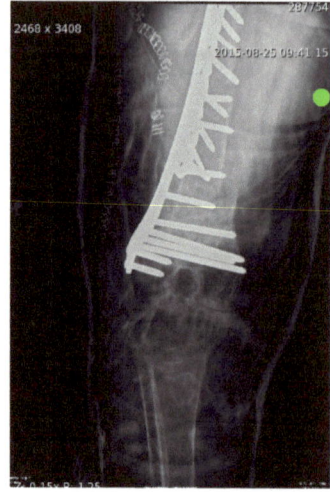

Fractures of the Patella

The patella is usually fractured from direct trauma or forceful knee extension while the knee is semi flexed during a fall. The knee should be assessed thoroughly especially the extensor component.

Radiographical images include AP, lateral, and sunrise views of the knee joint. Non displaced fractures should be treated conservatively in a knee immobilizer for 4-6 weeks. Displaced fractures on the other hand have to be managed surgically with tension band wires, cerclage wiring and screws.

Fractures of the Tibial plateau

The tibia comprises as the sole weight bearing bone in the lower leg through which body weight is transmitted. The tibial plateau articulates with the distal femur via the medial and lateral tibial plateaus. The 2 plateaus are separated by the intercondylar eminence where the anterior and posterior cruciate ligaments are attached. The tibial plateau serves as an insertion site for tenderness structures including the patellar ligament anteriorly, to the tibial tubercle, pesanserinus medially and attachment of the iliotibial band laterally. Important neurovascular structures lie anteriorly to the tibia namely the deep peroneal nerve and the popliteal artery, posteriorly.

Fractures of the tibial plateau usually occur due to a force applied laterally or medially to the tibia along with axial loading usually after high energy trauma from moto vehicle accidents and falls from lower heights.

Patient presents with swelling and pain around the knee. The deep and superficial peroneal nerves and the medial and lateral plantar nerves should be assessed along with the pulses of the lower limb. After the swelling has subsided the ligamentous structures of the knee should be tested.

Radiographical images include AP and lateral, oblique views of the knee also in 40 degrees internal and external rotation to assess the lateral and medial tibial plateaus. If the x-rays are not satisfactory a CT scan can be performed to assess the articular surfaces. MRI should be done to rule out ligamentous injuries. If lower limb pulses are diminished or not palpable, an arteriography should be done.

Schatzer Classification for Tibial Plateau Fractures

Type	Description	Surgical Management
I	Lateral plateau; split fracture; MCL injury	Passive Motion, Hinged cast brace, partial weight bearing; in displaced fractures ORIF with 2 lag screws or a buttress plate
II	Lateral plateau; split depression fracture	Skeletal Traction, ORIF with buttress plate and cannulated screws
III	Lateral plateau; depression fracture	ORIF with raft screws and bone graft
IV	Medial plateau fracture	Same as Type II
V	Bicondylar plateau fracture	External Fixator followed by definite internal fixation
VI	Plateau fracture with extension into metaphysis	Same as Type V

The patient presenting with this type of fracture should be immobilized with a knee brace in full extension and should ambulate with the help of crutches without applying weight on the affected limb. Initially the fracture should be reduced by skin traction until definitive management is decided. In fractures that are displaced, a posterior splint or a fixed internal fixator should be applied. Non-displaced fractures can be treated conservatively with minimum weight bearing, exercises and limited knee range motion in a knee brace. The fractures can be followed radiographically to catch any displacement. Full weight-bearing exercises may be initiated after 3 months.

Immediate surgical intervention should be considered in neurovascular injury, displaced fractures and compartment syndromes with external fixator and subsequent open reduction, internal fixation with nail plate and screws.

Shaft fractures of Tibia and Fibula

Fractures of the shaft of the tibia and fibula are due to rotational injuries in high or low energy motor vehicle accidents, usually together. A patient with such a fracture should be thoroughly assessed for neurovascular injuries, which include the superficial and deep peroneal, medial and lateral plantar nerves along with the posterior tibial and dorsalis pedis artery. Compartment syndrome should also be ruled in this type of fracture, due to the tight fascial compartments in the lower limb of the lower leg.

Radiographical images include AP and lateral x-rays of the tibia and fibula with the knee and ankle joints. Shaft fractures can be classified according to the general description of a fracture as mentioned above.

The fracture should be reduced immediately and stabilized in a long leg cast with the knee in 5-degree flexion. The patient should be advised clutches and minimal weight bearing on the effected leg. Displaced or comminuted fractures should be treated surgically with intramedullary nailing, external fixator and nail plate and screws. Shafts of the fibula is usually treated conservatively by immobilization in a walking cast.

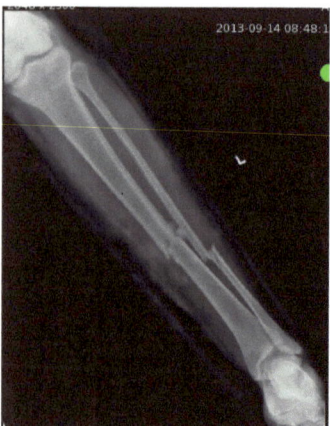
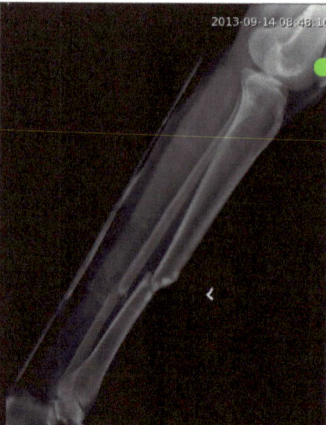
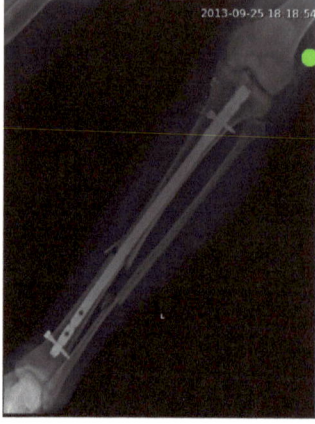

Ankle Fractures

The ankle joint is a complex – hinged type joint constructed by the tibia, fibula and articulating surfaces of the talus with a strong scaffold of ligaments placed medially and laterally. Patients with fracture of the ankle should have their neurovascular examinations performed thoroughly mainly assessing superficial and deep peroneal, medial and lateral plantar nerves, posterior tibial and dorsalis pedis artery.

The patient typically presents with pain swelling and difficulty with weight bearing on the effected limb after stumbling and advancing on a twisted anchored ankle.

Radiographical images include AP, lateral and 20-degree of internal rotation x-rays of the ankle. Additional radiographical images of the knee joint, tibia and fibula should be taken. Ligamentous injuries if suspected can be assessed by an MRI.

The Weber classification can help classify ankle fractures according to the level of the fibular injuries. This classification can go hand in hand with the Lauge- Hansen classification which classifies ankle fractures according to mechanism of injury which includes the direction of the force of the fracture and position of the foot.

Weber Classification/Lauge-Hansen Classification of Ankle Fractures

Weber A	Fibular fracture below tibial plafond	Supination adduction fracture	Medial displacement of talus and transverse fracture of distal fibula and/or vertical medial malleolus fracture
Weber B	Spiral/oblique fracture of fibular occurring near level of syndesmoses	Supination external rotation fracture	Disruption of anterior tibia fibular ligament; spiral fracture of distal fibula, posterior malleolar fracture and fracture of medial malleolus or injury to deltoid ligament
Weber C	Fracture above level of syndesmosis	Pronation abduction	Medial malleolar fracture, distal fibular fracture, deltoid ligament injury
Weber C	Fracture above level of syndesmosis	Pronation external rotation	Medial malleolar fracture, distal fibular fracture, deltoid ligament injury

The patient presenting with ankle fractures, should have immediate close reduction followed by a well patted posterior splint with stirrups. The limb should be elevated at the level of the chest to ensure adequate bone union. After reduction an x-ray should be taken to make sure that the talus and tibia are in a good position.

Non-displaced fractures overall can be treated conservatively and along with a cast for up to 2 months with minimal weight bearing. On the other hand, displaced fractures should be advisedsurgical management with nail plate and screws with our without syndesmotic screws.

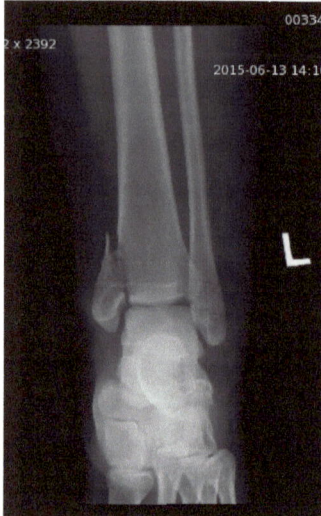 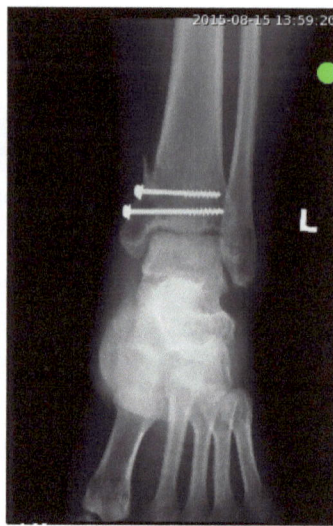

Pilon Fractures

The tibial Plafond also known as the 'pilon' is the articulating region between the distal tibia and the talus. Fracture of this region usually occurs after high energy motor vehicle accidents or falls resulting in compression of the tibial plafond between the talus and the tibia.

The patient typically presents with swelling and pain of the ankle. The neurovascular exam includes the superficial and deep peroneal, medial and lateral plantar, deep dorsalis pedis and posterior tibial artery should be assessed.

Radiographical image include AP, Lateral, 20-degree internal rotation x-rays of the ankle. A CT maybe used to enhance the fracture and the articulating surfaces.

Ruedi and Algower Classification

Type	Description
I	Non-displaced fracture
II	Displaced fracture with minimal impaction and comminution
III	Displaced fracture with significant comminution and metaphyseal impaction

Immediate reduction of this type of fracture should be performed along with placement of a splint to avoid skin necrosis and blistering. Non-operative management includes a long leg cast for 6 weeks with minimum weight bearing and range of motion exercises. Displaced fractures are managed surgically after swelling and soft tissue injury has subsided with an external fixator or open reduction internal fixation with plates and screws.

Fractures of the Talus

Injuries of this bone occurs from falls from high heights or high energy trauma from motor vehicle accidents that cause hyper-dorsiflexion of the ankle compressing the neck of the talus against the anterior edge of the tibia. The blood supply of the talus is minimally inadequate making it prone to avascular necrosis.

Patient typically presents with foot pain and swelling of the ankle and foot

Radiographical images include AP, lateral and 20-degree internal rotation x-rays of the ankle and AP lateral and oblique views of the foot. A canal view may also be taken with the ankle in planter flexion 15-degrees pronation with the radiograph machine in a 15-degrees from the vertical. This view allows an adequate identification of the talus neck. A CT scan may also be advised to delineate the articular surface and fracture. The talus may be effected in the neck, body, head, and lateral and posterior process.

Non-displaced fractures can be treated non-operatively with a short leg cast with non-weight bearing for 6 weeks. Displaced fractures should be treated surgically with open reduction and internal fixation with nail plates, screws, and k-wire.

Hawkins Classification of the Talus fracture

Type	Description
I	Non-displaced
II	Talus fracture plus Subtalar dislocation
III	Talus fracture plus Subtalar and tibiotalur dislocation
IV	Talus fracture plus Subtalar, tibiotalur and talonavicular dislocation

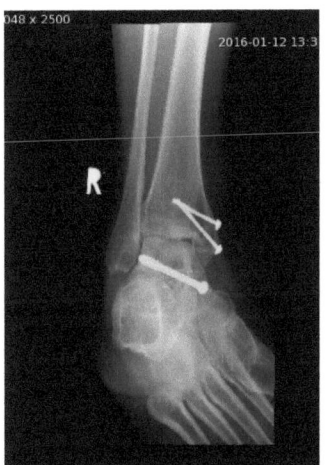

Calcaneal Fractures

These fracture usually occur in middle aged men from significant falls or motor vehicle accidents. Patients present with heal pain, swelling and bruising.

Radiographical images include lateral radiograph of the hind foot and AP view of the foot 'Haris axil view' and AP lateral 20-degrees internal rotation of the ankle. Maximal dorsiflexion is seen in the Haris axil view with the radio graph machine, 45-degress cephalad to identify articular ring surfaces. If the x-rays are inconclusive a CT maybe performed. The calcaneum maybe be fractured in various locations including the anterior process, calcaneal tuberosity, medial process, sustentaculum tali and body. Lateral radiographs can help assess anterior process and calcaneal tuberosity fractures. Axial CT scans can help assess fractures of medial process, sustentacular tali and body fractures.

Immediate management should be undertaken to prevent chronic pain. Non-operative management is usually reserved for non-displaced fractures with reduction and placement in a bully joint splint with elastic compression stockings and non-weight bearing of the effected limb. Displaced fractures are treated surgically with many fragment screws or lag screws with cerclage wire depending on the location of the fracture.

Navicular Bone Fracture

The navicular bone serves as the interconnection between the subtalar joint and the forefoot instituting the majority to the foots medial longitudinal arch. Fracture in this bone present with pain in the foot and dorsomedial swelling and tenderness. Radiographical images include medial and lateral oblique views of the midfoot, along with AP, and lateral views. The body of the navicular bone is usually effected upon which it is classified in the following table.

Type	Description
I	Body fracture dividing into dorsal and plantar pieces.
II	Navicular body fractures splitting into medial and lateral pieces.
III	Communicated navicular body fractures with medial and lateral pole displacement.

Fractures that are not displaced can be treated in a cast or boot with no weight bearing for 6-8 weeks. Disrupted fractures require surgical intervention with open reduction internal fixation with lag screws.

Tarsalmetatarsal Fractures (LisFranc Fractures)

Patients present with swelling and pain in the mid-foot region. The tarsalmetatarsal joint has the following articulations: Between the medial and lateral cuneiforms of the 2nd and 3rd metatarsal mainly stabilized by the 2nd metatarsal base and the LisFranc ligament spanning from the medial cuneiform to the base of the 2nd metatarsal.

Fractures in this region usually occur due to crush injuries, axial loading and forced abduction of the foot. The dorsalis pedis artery should be assessed as it travels between the 1st and 2nd metatarsal bones.

Radiographical images include AP, lateral, oblique views of the foot. Weight-bearing views should be performed to assess displacement. Non-displaced fractures are treated conservatively with non-weight bearing and rest. Displacement fractures are treated surgically with screws and K-wires. A CT view should be done to delineate the fracture.

Miscellaneous Fractures of the Lower limb

Fractures of the 1st 2nd 3rd and 4th metatarsals along with **fractures of the phalanges of the toes** usually occur due to crush or axial loading. AP lateral and oblique vies of this fracture should done. Treatment is typically conservative with hard-soled shoes with progressive weight bearing. Displaced fractures can be dealt surgically with screws and K-wire.

Cuboid bone fractures have a similar presentation with the navicular bone fractures. Non conservative management with boot non-weight-bearing and/or cast for 6-8 weeks. Disruption fractures require surgical intervention with open reduction internal fixation.

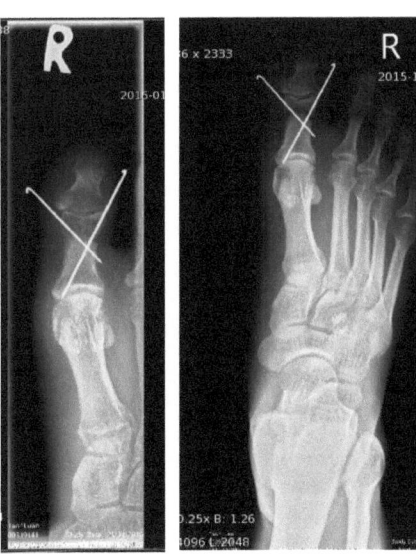

Dislocations

Hip Dislocation

Posterior hip dislocations are more common than anterior hip dislocations due to high energy trauma from motor vehicle accidents. These are usually associated with sciatic nerve injury while anterior hip dislocations are associated with femoral head injury. Posterior hip dislocations present with a shortened extremity with the hip flexed, internally rotated and adducted. Patients with anterior hip dislocation have the effected limb in external rotation, flexion and abduction. These patients should have a thorough neurovascular examination especially focusing on the peroneal aspect of the sciatic nerve due to its superficial location. Radiographical images include AP view of the pelvis and AP, and lateral views of the femur. Patients with hip dislocation should be reduced immediately to prevent avascular necrosis with the help of relaxation of the patients muscles with sedation. After reducing the dislocation, radiographs should be obtained along with the hip being assessed for subluxation by flexing the hip to 90-degrees and applying a posterior directed force. In cases of subluxation, surgery is indicated.

Knee Dislocation

Knee joint dislocation usually occurs due to disruption of the ligaments of the knee mainly the ACL, PCL and medial and lateral collateral ligaments. Knee dislocations are described according to the relationship of the proximal tibia to the distal femur for instance the i.e. anterior, posterior, lateral and medial rotation. Neurovascular examination should be performed where the popliteal artery and tibial and fibular nerve should be tested thoroughly as well. These dislocations should be reduced immediately following reduction. Radiographical images include AP, lateral and notch views of the knee and sunrise view of the patella should be taken along with the recheck of the neurovascular system. An MRI can be done to detect soft tissue or ligamentous injuries and arteriography of the lower limb in case of diminished lower limb peripheral pulses. After confirmation of proper reduction with the radiographical images, the knee should be placed in a splint in 30-degrees flexion. Surgery is advised for cases of failed closed reduction, soft tissue injury and vascular injuries.

Patella Dislocation

The patella can be dislocated due to high energy trauma in acute settings or in a recurrent fashion (chronic in patients with connective tissue disorders). Patients present with the knee in the extended positon unable to flex the knee, hemarthrosis, and grossly palpable patella in an abnormal position. Patients with recurrent patellar dislocations have a positive apprehension test. Radiographical image include AP, lateral, tangential (skyline) views of the knee and sunrise view of the patella. Patellar dislocations should be reduced followed by a cast or brace to the knee in extension.

Subtalar Dislocation

Subtalar dislocation is the loss of articulation between the talus and the calcaneus and navicular bone. Medial subtalar dislocation occurs due to inversion of the foot while lateral subtalar dislocation occurs in eversion. Subtalar dislocation should be reduced immediately if not possible, then surgery should be done.

Soft tissue injury

Tear of the Quadriceps Tendon

The main site of quadriceps tendon rupture tear is within 2 cm of the superior part of the patella. Quadriceps tendon rupture is associated with anabolic steroid use, local steroid injection, Diabetes mellitus, inflammatory arthropathy and chronic renal failure. Patients present with pain in the knee, difficulty bearing weights, knee joint effusion, tender upper pole of the patella and palpable defect at the superior pole of the patella. Radiographical images include AP, lateral and sunrise views of the knee. These patients should be treated nonoperatively with immobilization of the knee in extension in a plaster cylinder for 4-6 weeks followed by physical therapy. In cases of complete tear surgery consisting of a repair reinforced by a partial thickness quadriceps tendon flap is indicated.

Tear of the Patellar Ligament

As appose to the tear of the quadriceps tendon tear of the patellar ligament occurs in the inferior pole of the patella. These tears are associated with rheumatoid arthritis, SLE, diabetes mellitus, renal failure, corticosteroid treatment, local steroid injection and chronic patellar tendonitis. Patient gives a history of a 'pop' sound in the forceful knee extension on clinical examination a defect maybe palpated along with hemarthrosis, decreased range of motion and loss of active extension. Radiographical images include AP, lateral views of the knee to visualize a high riding patella. Surgical repair or reattachment to the bone, as soon as possible is advised in complete tears.

Ankle Sprain

Forceful eversion or inversion the ankle and foot may lead to an ankle sprain. This ankle sprain is diagnosed if no fracture or dislocations are found in radio images. The anterior talofibular and calcaneofibular ligaments are the most common ligaments injured. These patients are advised to rest, apply ice to the affected area, compression with elastic ace wrap and elevation of the foot along with pain relief and non-weight bearing.

Achilles Tendon Rupture

The Achilles tendon connects the gastrocnemius muscle to the calcaneus. Repetitive use of the calves may lead to a rupture of this tendon. On clinical examination a palpable defect maybe be felt above the heel. A Thompson test, plantar-flexion squeeze is positive or where no plantar flexion occurs when the calf is squeezed and the patient is also unable to perform heel raise. Nonoperative treatment includes a 2-week immobilization in a plantar flexed splint followed by immobilization in a cast for 8 weeks with progressive weight bearing. Surgical treatment is usually kept aside for athletes percutaneously through a medial longitudinal approach.

Peroneal Tendon Subluxation

This usually occurs in winter sports such as skiing. Patients present with lateral ankle swelling and tenderness in the posterior aspect of the lateral malleolus. The tendon should be reduced and the foot should be placed in a well moulded cast in plantar flexion and inversion.

INJURIES TO SOFT TISSUE OF THE KNEE JOINT

Patients with injuries to the soft tissues of the knee present with pain, swelling and decreased range of motion of the knee joint.

In cases of medial meniscus injury, the patient is unable to extend the knee with pain and tenderness in the anterior or posteromedial joint line. An MRI is done to make the diagnosis. The patient should be treated non-surgically with pain relief and alleviation of the swelling by NSAIDs and aspiration, respectively. Isometric quadriceps exercise with the knee in maximum extension should be advised. Arthroscopy is the surgical management to debride central tears or reconstruct/repair peripheral tears.

Medial collateral ligament injury is associated with rotational or valgus stress related injuries. The medial aspect of the knee is tender along with swelling. A knee brace in the varus position may be applied to relieve the medial collateral ligament if partially torn.

Lateral collateral ligament injury is associated with disruption of the popliteus muscle tendon, iliotibial band, head of fibula or common peroneal nerve. The patient perceives tenderness in the lateral aspect of the knee. A knee brace for partial tears and surgical reconstruction for complete tears is advised.

Tears of the anterior and posterior cruciate ligament result in an anterior and posterior displacement of the tibia in relation to the femur, respectively. This can be assessed clinically with a positive anterior and posterior drawer test. An MRI should be done to delineate the ligamentous injuries. Non-surgical treatment with a knee brace and full range of motion exercises is advised followed by reconstruction arthroscopically if necessary.

Upper Limb Fractures

Clavicular Fractures

The clavicle is an S shaped bone that connects the shoulder to the trunk. Fractures of this bone usually occur due to fall on the ipsilateral shoulder, fall on an outstretched hand and direct blow. The clavicle guards the important neurovascular supply to the upper limb mainly naming the brachial plexus, subclavian and axillary vessels, making it mandatory to rule out injuries to the neurovascular supply of the upper limb.

An AP view, and 30 degrees cephalic tilt of the chest is enough to make the diagnosis, although a CT may be used in cases to confirm a sternoclavicular dislocation.

Clavicle fractures are treated non-surgically with a sling for 6-8 weeks. Fractures that are open, associated with injury to the neurovascular supply of the upper limb, lateral one-third fractures and tenting of the skin should be dealt with surgery

Type	Description
I	Middle Third Fracture
II	Distal/Lateral Third Fracture a- interligamentous between the conoid and trapezoid ligament OR coracoclavicular and acromioclavicular ligament b- medial to coracoclavicular ligament OR between the conoid and trapezoid ligament with the conoid ligament torn c- No ligamentous Injury
III	Proximal/Medial Third Fracture

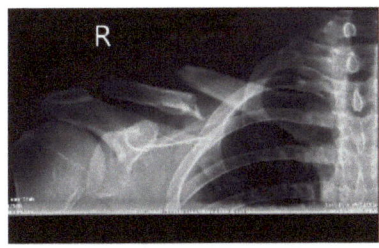

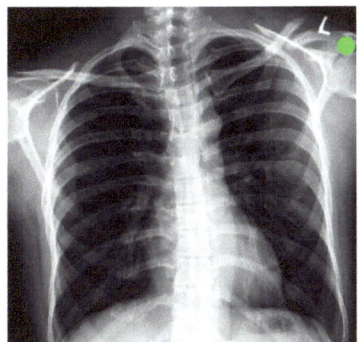

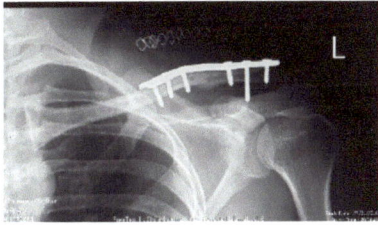

Fractures of the Scapula

Scapular fractures are classified according to the region affected, body, neck, spine, acromion, glenoid and coracoid. Due to its close proximity to important structures, subclavian vessels, aorta, lungs, ribs and brachial plexus, the body of the scapula is associated with injuries related to these structures most commonly with high energy trauma. The articular surface must be carefully evaluated of the glenoid to ensure that there is no shoulder instability when there is a blow to the shoulder or a fall on the outstretched arm.

Radiographical images include AP view in the plane of the scapula, axillary x-ray assisted by an axial view of the scapular body and transscauplar views. A CT scan may be required if surgical intervention is needed.

Non-operative management includes a sling for 4-6 weeks. Certain criteria must be met in order to proceed with the surgical management which include displaced intra-articular fracture (more than 25% of the articular surface), scapular neck fracture (more than 40% angulation and/or 1 cm of medial translation), scapular neck fractures with associated displaced clavicular fracture, acromion fractures and coracoid fractures.

Proximal Humeral Fracture

Elderly individuals with osteoporosis resulting in a fall are vulnerable to fractures of the proximal humerus, including the humeral head and neck. Injuries associated with fractures in this region should be assessed for neurovascular injuries, dislocation, rotator cuff tears. Nerve sensation testing of the axillary, lateral aspect of the shoulder and the overlying deltoid should be done.

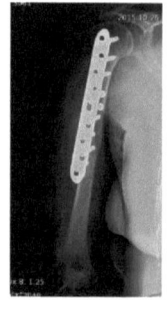

Radiographical images include AP, lateral, scapular Y and axillary views. Avelpeau axillary view should be performed if the patient is not capable of performing the standard axillary view, in a sling leaned obliquely backward 45 degrees over the cassette with the beam directed caudally. A CT scan may be done to evaluate further associated fractures.

Immediate management should be attained in order to prevent long term complications such as avascular necrosis of the humeral head due to disruption of the arcuate branch off of the anterior circumflex humeral artery.

Neer's Classification
Parts of the Proximal Humerus involved
Head
Shaft
Greater Tuberosity

Minimally Displaced	Displaced
Less than 1 cm and <45 degrees of angulation which can be treated in a sling with an early gentle range of motion exercises	More than 1 cm or >45 degrees of angulation, superior displacement of the greater tuberosity fragment of > 5mm, lesser tuberosity fractures that block internal rotation usually requires surgery: Closed reduction and percutaneous fixation, open reduction

	and internal fixation and prosthetic arthroplasty

Fractures of Humeral Shaft

According to its functional region humerus fractures can described: open vs closed, location (proximal, middle, distal third), non-displaced vs displaced. Direct trauma or falls can result in these type of fractures. The proximal segment may be adducted by the pectoralis major in proximal shaft fractures while the proximal segment is abducted by the deltoid in distal shaft fractures. A thorough neurovascular examination should be performed keeping mind of the radial neve.

Radiographical images include AP, and lateral views of the humerus along with the shoulder and elbow joint to oversee any associated injuries.

Non-operative management, which is the safest and cheapest option includes a cast, splint or brace. A protective functional brace and a collar and cuff may also be used. Surgical management may be advised in failed non-operative management, open fractures, concomitant vascular injury, 'floating elbow', concomitant fracture of the forearm bones. The most common injury of the middle third fracture of the humerus includes the radial nerve due to which open reduction with internal fixation with nail plate and screws is superior to intramedullary nailing to further explore the neurovascular injuries.

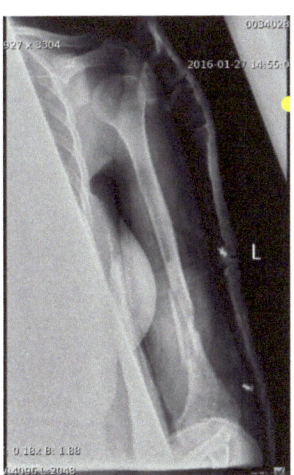
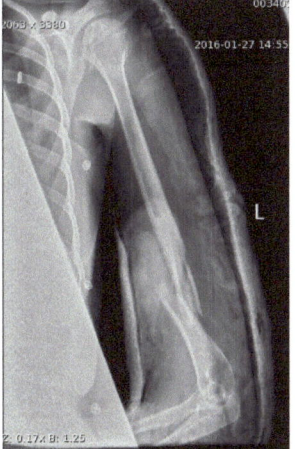
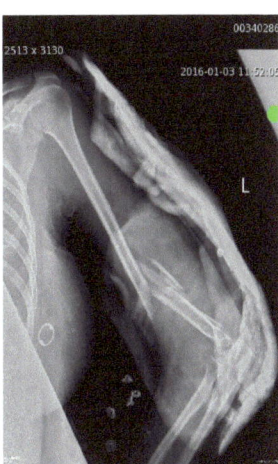
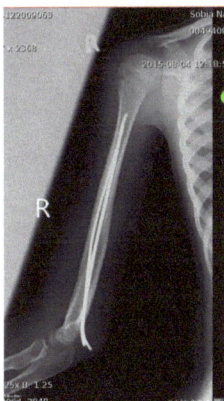

Distal Humerus Fractures

The distal humerus is composed of 2 condyles upon which fractures are classified according to the region affected: supracondylar, transcondylar, intercondylar, condylar, capitulum, trochlear, epicondylar, and supracondylar process fractures.

AO Classification based on column integrity and articular involvement

Type	Description
A	Extra articular; epicondylar, supracondylar, transcondylar
B	Involvement of a portion of articular surface; unicondylar, intercondylar
C	Involvement of entire distal articular surface

Radiographical images include AP lateral and oblique along with traction radiographs. ACT scan may be done before surgery. A non-displaced distal humerus fracture maybe recognized by a quote on quote fat pad sign on a lateral radiograph depicting displacement of the adipose layer of the joint capsule. The ulnar nerve should be kept assessed from time to time as it is at risk for injury. Patients with the distal humeral fractures can be treated conservatively with a posterior long arm splint with the elbow flexed at 90-degerees and the forearm in neutral position. This type of treatment is reserved for non-displaced fractures while displaced fractures with vascular injuries should be dealt with surgical intervention with open reduction internal fixation with nail plate and screws or total elbow arthroplasty.

Supracondylar fractures: Can be of 2 types

1) Extension where the distal fragment is displaced posteriorly and

2) flexion where the distal fragment is displaced anteriorly

Transcondylar Fractures:

Intercondylar Fractures: These fractures are the most common type of distal humerus fractures. They are prone to displacement due to muscle forces on the medial and lateral epicondyles that work in opposite directions with the muscles attached to the medial condyles causing flexion and muscles attached to the epicondyle causing extension.

Type	Description
I	Non-displaced
II	Slight displacement with no rotation between the condylar fragments
III	Displacement with rotation
IV	Comminution of the articular surface

Condylar Fractures: The medial or lateral condyles are usually effected in these type of fractures

Milch Classification	

Type	Description
I	No lateral trochlear ridge involvement
II	Involvement of lateral trochlear ridge with medial-lateral instability due to which while placing a posterior long arm splint the forearm should be supinated for lateral condyle fractures or pronated for medial epicondyle fractures with the elbow flexed at 90-degrees.

Capitulum Fractures: These fractures lead to free articular fragment that is displaced anteriorly into the coronoid or radial fossa causing the inability to flex the elbow.

Type	Description
I	(Hahn-Steinthal) large osseous fragment with or without trochlear involvement
II	(Kocher-Lorenz) fragment articular cartilage with minimal subchondral bone attached.
III	Significant comminution

Trochlear fractures

Epicondyle fractures:

Supracondylar process fractures: the supracondylar process is an osseous projection of the anterior medial surface of the humerus

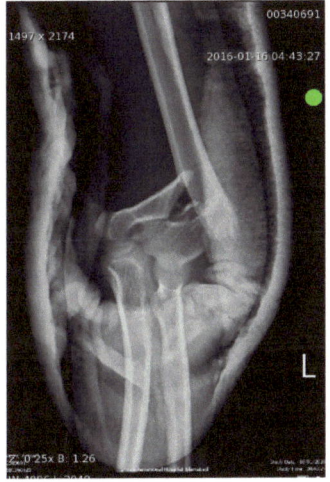
the

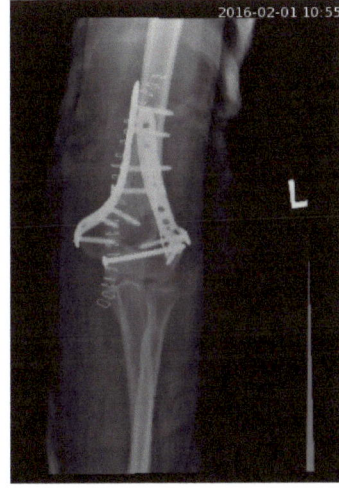

Fractures of Proximal Ulna

Olecranon Fractures

Since the olecranon and its position of the subcutaneous tissue along with its proximal location on the ulna, it is highly susceptible to direct trauma. Active elbow extension may be lost due to displaced fracture of the olecranon involving the triceps function. Stability of the elbow joint is provided by the interior portion of the ulna which is the coronoid process. A direct blow or fall on an outstretched arm may result in a comminuted, transverse, oblique fracture of the olecranon. These fractures should be preceded neurovascular examination.

Radiographical images include AP and lateral views keeping in mind of the displacement of the radial head. On radiograph, radial head dislocation is present when the radial head does not point towards the capitulum in all views.

Management of non-displaced fractures should be advised with closed treatment in a long arm splint or a cast with the elbow flexed from 45-90 degrees. Progression of extension and flexion are sufficiently stable from 3-6 weeks. Surgical management should be done in any case of disruption of the extensor mechanism or articular incongruity. Surgical options include intramedullary fixation, tension band wiring, plate and screws and excision

Mayo classification of olecranon fractures	

Type	Description
I	Non-displaced
II	Displacement without elbow instability
III	Fracture with features of elbow instability

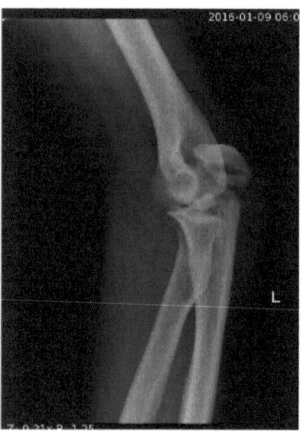
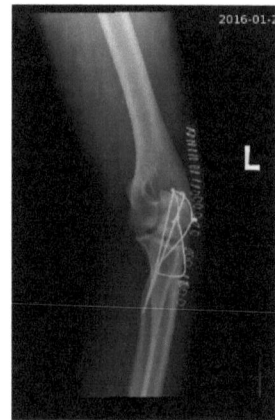
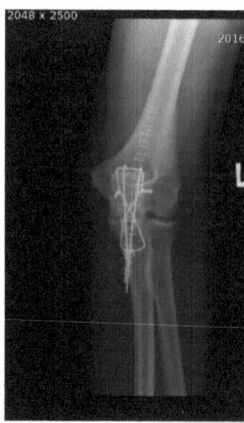

Coronoid Fractures

The attachment of the MCL along with the anterior capsule allow elbow stability of the coronoid process. Posterior elbow dislocation is usually associated with coronoid fractures. Posterior displacement of the proximal ulna or hyper extension force of the elbow usually result in these type of fractures.

Radiographical images include the oblique view of the elbow.

Regan and Morrey Classification based on the size of the fracture fragment

Type	Description	Treatment
I	Coronoid Process tip involved	Immobilization in flexion for 3 weeks
II	Single or comminuted fragment involving 50% or less of the coronoid process	Operative Intervention if unstable
III	A single or comminuted fragment involving more than 50 % of the process	Operative Intervention if unstable

Fractures of the proximal radius

Radial head fractures

Falls usually result in radial head fractures due to the radial head and capitulum colliding against each other. Range of motion is diminished along with pain in these fractures. An essex-lopresti type (radial head fracture dislocation with associated interosseous ligament and distal radioulnar joint disruption) injury should be looked out for if tenderness is positive throughout the neurovascular examination of this region.

Radiographical images include AP, lateral, radial head views. If a fat Pad sign is positive, a non-displaced fracture can be diagnosed.

In order to prevent future block to motion, aspiration of the hemarthrosis along with injection of lidocaine should be performed to assess the valgus stress.

Non-operative management of radial head fractures is preceded by immobilisation in a sling followed by early range of motion 24-48 hours after injury. Type III fractures and block to motion are the only indicators for surgical intervention. Fragment excision with or without prosthetic replacement and open reduction and internal fixation are surgical options along with closed reduction specifically for Type IV injuries.

Mason Classification

Type	Description	Treatment
I	Non-displaced vertical split fracture	Collar and Cuff Placement
II	Marginal fractures with displacement	ORIF with 2 small headless screws
III	Comminuted Fracture involving the entire radial head	ORIF with small headless crews or metal spacer
IV	Fracture associated with elbow dislocation	Prosthetic Replacement

Fractures of the forearm

The radial and ulnar shaft fracture are more common in men due to high energy trauma, athletic injury and falls. Any disruption in the radius and ulna will cause instability of the radio-ulnar joint. Stability is achieved by the interosseous membrane which occupies the space between the radius and ulna. Due to the tight compartments present in the forearm, compartment syndrome along with open wounds should be assessed in a thorough neurovascular examination.

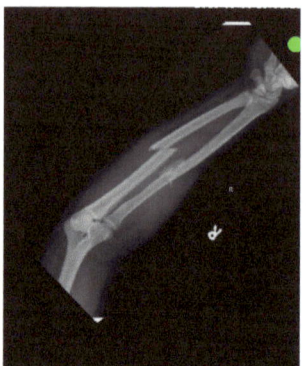
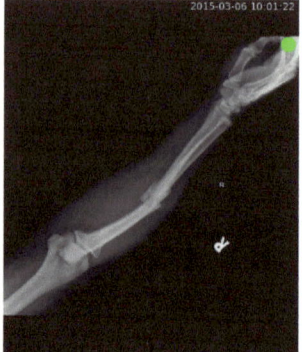
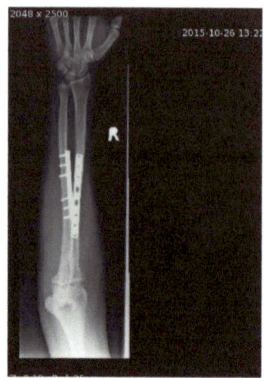

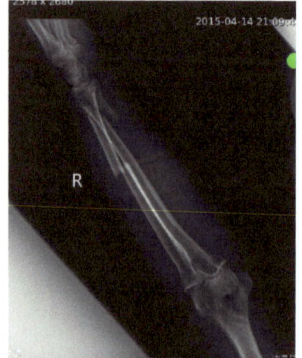
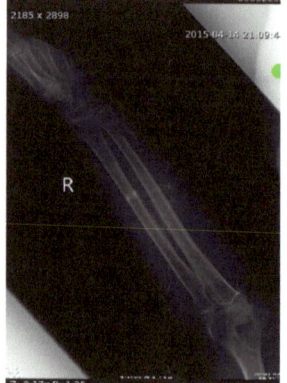
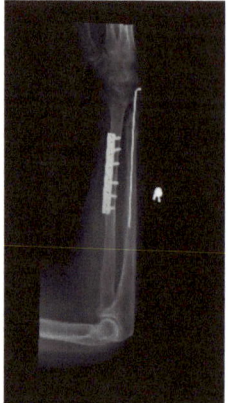

Isolated radial shaft fractures

Patients with radial shaft fractures are usually victims of direct trauma or falls on an outstretched hand. Fractures of the distal third of the radius is usually associated with disruption of the distal radioulnar joint and these patients present with wrist pain. Radiographical images include AP, and lateral views of the forearm, wrist and elbow joint. Disruption of the distal radioulnar joint is signified by fracture at the base off the ulnar styloid widening of the distal radioulnar joint space, subluxation of the ulna and radial shortening >5mm to the distal ulna. Patients without displacement should be treated conservatively with a long arm cast while surgical intervention is kept aside for displaced fractures with disruption of the distal radioulnar joint.

Surgical options include open reduction and internal fixation with 3.5mm DCP plates along with transfixing pins and K-wires.

Isolated Ulnar Fractures

These fractures are also known as night stick fractures owing to the night sticks used by our fellow policemen. As the name states these fractures occur due to a direct blow to the ulna. Neurovascular examination should be performed thoroughly to rule out any injury to the neurovascular supply to the forearm and hand.
Radiographical images include AP, and lateral views of the forearm along with the wrist and elbow joint. Fractures that are non-displaced can be treated with a sugar tong splint or a long-arm cast. Fractures that are displaced with >10-degrees angulation or greater than 50% displacement of the shaft are treated surgically with open reduction and internal fixation.

Monteggia's Fractures

Fracture of the proximal ulna with the radial head dislocation is known as 'Monteggia's fracture'. As all forearm fractures, neurovascular examination should be performed specifically keeping in mind the radial nerve and the posterior interosseous nerve. The patient presents with an obvious ulnar deformity with pain on the lateral aspect of the elbow. Radiographical images that should be taken include an AP, and lateral views of the elbow.

Bado Classification based on the direction of the radial head dislocation	
Type	Description
I	Anterior
II	Posterior
III	Lateral or Anterior/lateral
IV	Anterior dislocation with the fracture of the radius and ulna

All patients should be referred for surgical intervention with open reduction and internal fixation with plates and screws. Fixation of the ulna ultimately reduces the radial head.

Galeazzi Fractures

Galeazzi fracture is composed of a distal radial fracture with subluxation/dislocation of the inferior radio-ulnar joint usually after a fall on the hand. In addition to the presence of tenderness over the distal ulna, the piano key sign is positive indicating instability of the radioulnar joint by balloting the distal ulna. Radiographical images include an AP, and lateral views of the wrist and forearm. Surgical management of open reduction and internal fixation of the radial fracture with a compression plate and screws.

Radioulnar Forearm fractures

Fractures of both the firearm bones require high energy impaction from motor vehicle accidents or falls from heights. These fractures are most of the displaced due to which thorough neurovascular examination and pressure assessment of the forearm to look for compartment syndrome should be done. Radiographical images include AP, and lateral views of the forearm, wrist and elbow. These fractures should also be dealt surgically with open reduction internal fixation with compression plating using 3.5 mm dynamic plates but an external fixator maybe applied in case of open fracture and soft tissue injury.

Distal Radius Fracture

Distal radial fractures occur due to falls on the outstretched hand or high energy trauma from motor vehicle accidents or falls from heights. The patients would present with a swollen, bruised and tender wrist. These fractures are usually displaced with the dorsal displacement in colles fracture (dinner fork deformity) and volar displacement in smith type fracture (garden spade deformity). Like all forearm fractures, neurovascular examination should be tested especially the median, ulnar and radial nerves.

Radiographical images include PA and lateral views of the wrist, elbow and shoulder as well.

Frykman Classification based on degree of articular involvement and concomitant fracture of the distal ulna

AO/Asif Classification of distal radius fractures

Type	Description
A Extra-articular Fractures	1. Isolated distal ulnar fracture 2. Simple radius fracture 3. Radial fracture with metaphyseal impaction
B Intra-articular complex fracture	1. Radial styloid fracture 2. dorsal rim fracture 3. Volar rim fracture
C. Intra-articular complex fracture	1. Metaphyseal fracture with radiocarpal congruity preserved 2. Articular displacement 3. Diaphyseal-metaphyseal involvement

The fracture should immediately be reduced by closed reduction to prevent further swelling and to provide pain relief along with median nerve decompression followed by a sugar tong splint replacement. In cases of displaced fractures and where non-operative treatment is refractory, surgical management should be imitated with closed reduction and percutaneous pinning with k-wires, internal fixator or open reduction internal fixation with volar locking plate and screws.

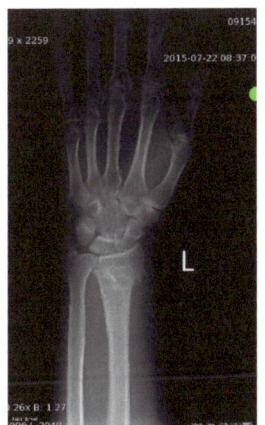 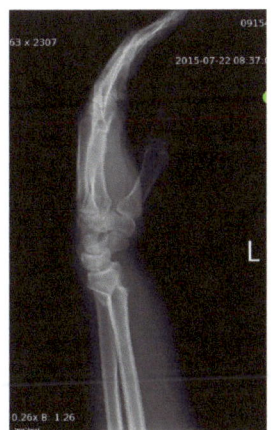 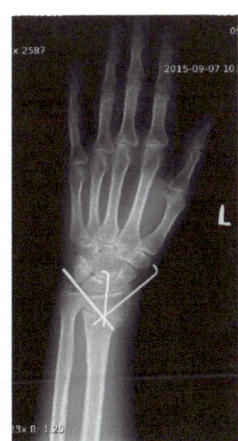

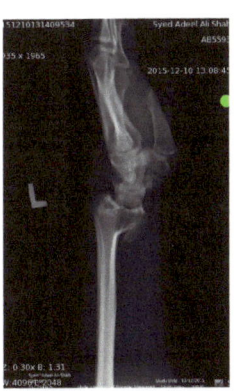 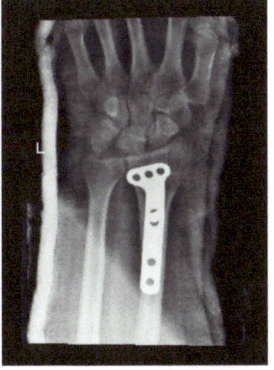 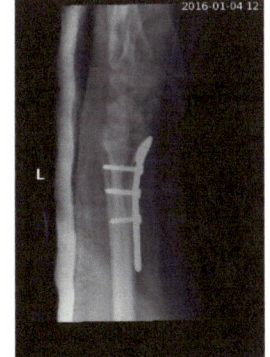

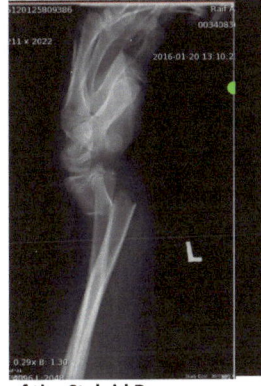 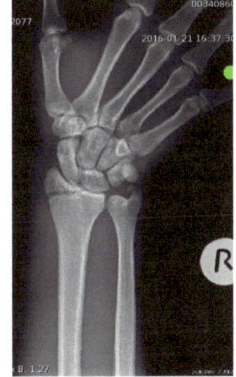 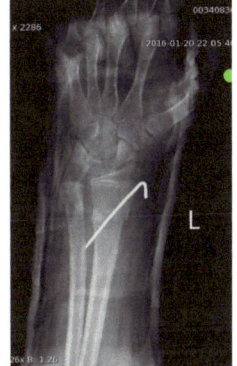

Fractures of the Styloid Process

Radial styloid fractures are also known as 'Chauffeurs fractures', 'Hutchingson fracture' and 'Backfire fracture'. It is characterized as an avulsion fracture with the extrinsic ligaments attached to the styloid ligament along with scapho-lunate dislocation or peri-lunate dislocation. The fracture is reduced in closed fashion followed by plaster slab from below the elbow to the metacarpal neck with the wrist held in ulnar deviation. Ulnar-styloid fractures usually occur with distal-radial fractures. Large ulnar styloid fractures can lead to a distal radioulnar joint instability that should be treated with open reduction and internal fixation.

Fracture of the carpal bones

From all the carpal bones the scaphoid bone is the most commonly fractured. Fractures of these bones occur to a fall on an outstretched hand these patients complain of pain on the radial side of the wrist and tenderness of the anatomical snuffbox region. The following tests should be perfumed in the physical examination: scaphoid lift test where the pain reproduced in dorsal volar shifting of the scaphoid and the Watson test where painful dorso-scaphoid displacement is reproduced as the wrist is moved from ulnar to radial deviation with compression of the tuberosity.

Radiographical images should include a specific scaphoid view which is an AP radiograph with the wrist supinated 30-degrees in ulnar deviation and a clenched fist view to assess carpal instability. X-rays may be insensitive to delineate scaphoid fractures so a CT scan may be advised in such cases. Non-displaced fractures are placed in a long arm thumb SPICA cast for 6 weeks. The scaphoid bone is prone to avascular necrosis at the waist. Displaced fractures of >1mm radio-lunate angle, >15-degress, scapho-lunate angle >60-degress, hump back deformity or non-union after conservative treatment should be dealt surgically with a compression screw.

Fracture of the lunate

Out of all the carpal bones, the lunate is the most commonly to dislocate but fractures may occur due to fall on an outstretched hand. These patients have tenderness over the volar wrist over the distal radius and lunate with painful range of movement.

X-rays are usually inconclusive so a CT maybe required to make the diagnosis. Non-displaced fractures can be placed in a short-arm cast while displaced or angulated fractures are treated surgically. Osteonecrosis of the scaphoid and lunate known as 'Keinbock's Disease' can lead to radiocarpal degeneration making management difficult.

Fracture of the Hamate

Fracture of this bone usually occurs from a direct blow while swinging a baseball bat or a golf club with a sudden stop in motion. These patients present pain in the ulnar side of the hand over the hamate. Radiographical images include a carpal tunnel view at 20-degress supination oblique view of the wrist should be done. Most of the time radiographical images are inconclusive so a CT scan should be performed. Non displaced fractures are treated in a short arm cast for 6 weeks while displaced fractures are openly reduced an internally fixed with screw or k-wires.

Metacarpal fractures

Metacarpal fractures can occur in the head, neck, shaft and base. Fractures of the metacarpal head require immediate reduction to avoid post traumatic arthritis and joint incongruity. Stable fractures are reduced and splinted in the intrinsic plus position where the wrist is in slight extension. Radiographical images include AP, lateral and oblique views of the effected hand. The metacarpal phalangeal joint, flexed 60-90 degrees and the proximal and interphalangeal joint is extended.

Metacarpal neck fractures usually occur in fist fractures and boxers most commonly in the fifth metacarpal. These fractures should be reduced in the closed fashion. Unstable fractures should be treated surgically by open reduction and internal fixation or percutaneous pinning.

Metacarpal shaft fractures which are no-displaced can be reduced in splinting in the intrinsic plus position as mentioned above. Surgery is indicated for cases with dorsal angulation >10-degrees for the 2nd and 3rd metacarpal and >40-degrees for 4th and 5th metacarpal along with rotational deformity.

Thumb metacarpal base fractures may be extra or intra-articular.

Intra-articular fracture of thumb metacarpal base

Type	Description
I	Benetts fracture. Single fracture line separates the majority of the metacarpal from the volar lip fragment
II	Rolondo's fracture: Comminuted with a Y or T-pattern including dorsal and palmar fragments.

These 2 fractures are treated surgically with open reduction internal fixation or closed reduction with percutaneous pinning. Fractures of the base of the 3rd and 4th metacarpal are treated with splinting in the intrinsic plus position with early range of motion. Dislocation of the 5th metacarpal base and hamate bone also known as 'Reverse Bennetts fracture' is treated surgically as well.

Phalangeal Fractures

Phalangeal fractures can occur in the proximal distal and

Intra-articular fractures can be classified as condylar fractures of fracture dislocations

Unicondylar	Closed reduction +/- ORIF
Bicondylar	Closed reduction +/- ORIF
Osteochondral	Closed reduction +/- ORIF

Volar lip Fracture	ORIF if >1mm of displacement
Dorsal Lip fracture	ORIF if >1mm of displacement

Extra-articular fractures are treated with closed reduction with a fingertrap traction and splinting.

Distal phalangeal fractures specifically intra-articular dorsal lip fractures usually occur with an extensor tendon disruption causing a mallet finger. These patient should be treated with an extension splint for 6-8 weeks or closed reduction with percutaneous pinning.

Intra-articular volar lip fracture can occur with extensor digitorum profundus rupture causing a jersey finger more commonly seen in the ring finger in rugby or football players. It is treated surgically.

Dislocation

Acromioclavicular Joint Dislocation

The medial acromion and the lateral end of the clavicle hold articular surfaces for the acromioclavicular joint which is covered with fibrocartilage. Vertical and horizontal stability of this joint is provided by the AC ligament and the coracoclavicular ligament. Acromioclavicular dislocation usually occurs due to high energy trauma or a fall. Along with the radiographical images that include AP, scapular –Y, axillary views, a neurovascular examination should be performed. Besides the standard radiographs a stress radiograph in which weights are attached to the wrists should be taken of both shoulders to compare and contrast the distances of the coracoclavicular ligament. This is also to differentiate between various grades of AC separations. Non-operative management include a sling for Type I, II, and III for 4 weeks but surgical intervention may be performed on athletes with Type III injury. Types IV, V, VI require surgical intervention.

Type	Description
I	Strain of AC ligament
II	Rupture of AC ligament and strain of CC ligament along with slight superior displacement of the superior clavicle
III	Rupture of both AC and CC ligaments causing superior migration of the lateral end of the clavicle resulting in a palpable step.
IV,V,VI	Detachment of the deltoid and trapezius from the distal clavicle along with disruption of the AC and CC ligaments with posterior(IV), superior(V) and inferior(VI) displacement of the clavicle.

Sternoclavicular Joint Dislocation

As discussed above with the AC dislocation this dislocation although rare occurs due to the same mechanism. A bump on the medial end of the clavicle is visible on inspection. CT is the choice of test in this dislocation due to its ability to distinguish fractures of the medial clavicle from the SC dislocation but standard physical examination along with radiographical images including AP and AP cephalic tilt x-rays may also be done. Posterior dislocation which is less common can be proceed to more complications due to its close proximity to the thoracic structures, hence requiring emergent reduction if vascular compression is positive by shoulder retraction and a towel clip. Non-operative treatment includes an ice pack for the first 24 hours and immobilization with a sling or a figure of eight bandage.

Dislocation of the shoulder

The glenohumeral joint is vulnerable to dislocation due to its attachment in multiple planes along with its vast range of mobility. These joint, being the most common dislocated joint of the body also carries negative forces that balance out with its free range including labrum, negative pressure of the joint the glenohumeral ligaments and the compression force of the humeral head against the glenoid. Trauma is the major cause of this dislocation in various positions (abduction, extension, external rotation).

Patient with anterior dislocation have their arm externally rotated and abducted with an emptiness in the posterior shoulder area and fullness in the anterior shoulder area. Radiographical images include AP and axillary views of the shoulder. Proper neurovascular examination should be down especially of the axillary nerve due to its vulnerability in this type of dislocation. Attention to 2 discrete lesions should be kept in mind, Bankart lesion where tear of the capsule and avulsion of the glenoid labrum accompanies the dislocation and Hill-Sachs lesion where the posterolateral aspect of the humoral head is indented by the anterior edge of the glenoid socket.

Patients with posterior dislocations have fullness in the posterior shoulder area, emptiness in the anterior shoulder area, prominent coracoid process and restricted external rotation of the arm. Radiographical images include AP view of the shoulder and axillary view and AP view in the scapular plane.

Shoulder dislocations should immediately be reduced in a closed fashion by gentle traction in the prone position and traction-counter traction with the sheet (Hippocratic Method). Prerequisites for reduction include analgesia and relaxation of the patient. After reduction, x-rays and neurovascular examination should be performed followed by a sling placement for 4 weeks. Surgical intervention is left for anterior traumatic recurrent instability.

Dislocation of the elbow

Elbow dislocations are the result of fall on the outstretched hand. Radiographical images include AP, lateral views of the elbow. Elbow dislocations are described according to the relationship of the ulna to the humerus (posterior-lateral, posterior-medial, lateral-medial and anterior). This type of dislocation should be reduced in a closed fashion with the patient under sedation. After reduction neurovascular examination and radiographs should be performed followed by a posterior splint placement at 90-degrees flexion. Surgery is indicated in unsuccessful closed reductions 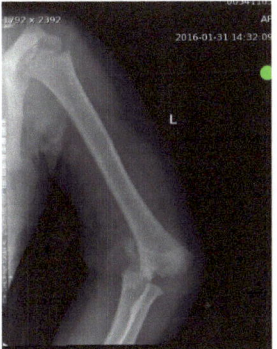 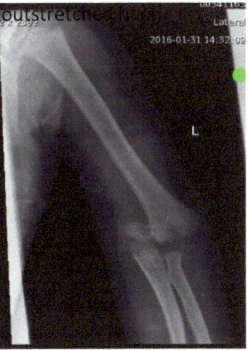 and recurrent dislocations. Elbow dislocations with radial head and coronoid process fractures is known a terrible triad, hence radiographical images should closely be monitored for concomitant fractures.

Dislocation of the digits

Much isn't said about the carpometacarpal dislocations besides the fact that they should be assessed by neurovascular examination and radiographical imaging along with immediate surgical intervention.

MCP joint dislocations maybe be simple or complex with both presenting in a hyperextended manner. Simple MCP dislocations may be reduced by flexion without traction due to its tendency to transition into a complex fracture whereas dislocation with the complex type with volar plates interposed are irreducible hence requiring surgery. A 'sesamoid' in the joint space is seen on radiographical images.

Due to its multiplanar motion, the MCP of the thumb is slightly more vulnerable and tends to sublux volarly around the opposite intact ligament. The most commonly injured ligament is the ulnar collateral ligament of the MP joint resulting in a Stener lesion which requires surgical intervention due to its inability to return to its original site. Skiers thumb is known as gamekeepers thumb an acute injury of the thumb metacarpophalangeal joint. Reduction is advised in these cases.

PIP dislocations include dorsal dislocation, pure volar dislocation and rotatory volar dislocation and these dislocations are reduced. Extension block splinting is advised for patients with dorsal dislocations which continue to sublux on lateral radiograph. Surgical intervention is required if failed reducible treatment.

DIP dislocations and thumb IP joint dislocations present late. Immobilisation in 30-degrees flexion for 3 weeks is required in unstable dislocations whereas Kirschner wire fixation is done in recurrent stability.

Soft Tissue Injury

Rotator Cuff Tears

Tear in the rotator cuff (supraspinatus, infraspinatus teres minor and subscapularis muscles) leads to proximal migration of the humeral head and subsequent rotator cuff arthropathy. These tears happen due to trauma, fall on the outstretched hand, overuse of the shoulder and degeneration of rotator cuff in the elderly. These patients present with shoulder pain and weakness in range of motion of the shoulder. On clinical examination the patient will have difficulty in overhead movements of the shoulder. These patients should be advised pain relief and physical therapy. In some cases, subacromial injection may be administered to control the pain and inflammation. Surgical repair can also be done for this type of injury.

Open Fractures

An open fracture also known as a compound fracture is a fracture in a bone that communicates with the external world by breaking through the overlying skin and soft tissue. These type of fractures complicate the management because of high rate of infection and injury to multiple structures such as the nearby muscles, tendons, ligaments and neurovascular bundle.

The open wound and surrounding structures should be thoroughly evaluated along with the neurovascular examination. In case of active bleeding direct pressure should be applied as opposed to tourniquet placement or instrumentation. After securing haemostasis the wound should be thoroughly irrigated with normal saline and covered with a gauze piece soaked in normal saline.

Radiographical images of the affected limb should be taken along with the proximal and distal joints. The limb may be reduced and splinted if ultimately necessary followed by a neurovascular check. If the vasculature is compromised or if the peripheral pulses are not palpable an arteriogram of the limb may be performed.

All patients should receive a tetanus shot and antibiotics prophylactically. Surgery should be conducted as soon as possible to avoid osteomyelitis by serial debridement of the fractured bone and surrounding soft tissue until all infected necrotic debris is removed followed by an external or internal fixator with bone grafting. A prophylactic fasciotomy may be done to prevent compartment syndrome.

Gustilo Anderson Classification

Type	Description	Antibiotic Treatment
I	Clean skin opening <1 cm	1st generation cephalosporin
II	Laceration of skin 1- 10 cm	1st generation cephalosporin
III	A- Extensive soft tissue injury B- Extensive soft tissue injury with periosteal stripping or bone exposure requiring a flap C- Associated Vascular injury	Ceftriaxone

Compartment Syndrome

An increase in pressure in a closed fascial compartment of a limb leading to impairment in vasculature is called compartment syndrome. The pressure may exceed to an extent causing muscle and nerve damage resulting in a debilitated limb without function. Compartment syndrome can occur in the arm, forearm, thigh, leg and foot.

The patient presents acutely with painful, cold, paresthetic, pale, pulseless and swollen limb with the pain aggravated by passive flexion of the extremity. The limb pressure should be assessed by a handheld pressure device. A pressure exceeding 30 mm Hg should be deployed with an emergent fasciotomy of the limb incising the skin and fascia followed by serial pressure monitoring of the limb. The other compartments of the limb may be released as well. The open wound should be covered with a sterile dressing or a vacuum assisted closure.

Infections

Osteomyelitis

Osteomyelitis is the infection of the bone and marrow. Patients present with bone pain, fever and chills. Radiographical images are usually unsatisfactory early on, due to which an MRI should be ordered. These images show a periosteal elevation with lytic lesions. Initially patients are started on antibiotic therapy followed by surgical debridement if conservative management is refractory. Surgical debridement also includes a culture and biopsy of the necrotic debris for specific antibiotic therapy intravenously. In cases where hardware is present, it should be removed.

Septic Arthritis

Septic arthritis is usually due to direct trauma or haematogenous spread from another source like osteomyelitis. Streptococcus pyogenes and staphylococcus aureus are the culprits for this infection. The patients present with pain in the affected joint with a pseudo-paralytic limb. A complete blood count shows leucocytosis, the CRP and ESR are elevated. Arthrocentesis reveals joint fluid with a WBC count >75000, and a positive culture. Radiographical images or an ultrasound may be performed to reveal joint effusion with capsular distention. An emergent drainage with intravenous culture directed antibiotic therapy should be commenced.

Vertebral Column

Cervical Fractures

Atlas

Jefferson Burst Fracture

Due to vertical compression and/or excessive load on the axis while the cervical spine is in the neutral position, neither extension nor flexion, may result in a stable fracture of the atlas. As the force is transmitted through the occipital condyles to the spine, the atlas is compressed between the occipital condyles and the axis resulting in the following two types of fractures:

1- Unilateral fracture of the lateral mass either anterior or posterior to it leading to a subsequent asymmetric displacement of the lateral mass.

2- Bilateral fracture in the anterior and posterior arches making a total of four fractures. This constitutes as a true Jefferson fracture. The intact ligamentous structures keep the fracture stable.

Common complaints of the patients include upper neck and vertex of the skull pain. There may be a laceration or hematoma on the vertex where the initial impact took place.

The best radiological study is an open-mouth view that visualizes the relation of the lateral masses to the underlying articulate surface of the axis. By measuring the degree of displacement on both sides the sum may be calculated yielding the total displacement. A value more than 5.7 mm creates a high suspicion of an unstable fracture but most of the time a value less than 5.7 mm indicates a stable fracture due to the intact transverse ligament.

After ruling other cervical fracture, the patient may be stabilized with a halo traction.

Axis

Hangman's Fracture

Bilateral fractures of the pars interarticularis of C2 with or without a C2/3 facet dislocation is known as a hangman's fracture. Extension with distraction is the mechanism of this injury, that occurs in motor vehicle accidents involving a head strike to the dashboard. Radiographical images include a flexion-extension x-ray of the cervical spine in the lateral view to determine the presence of displacement.

Undisplaced fractures are treated conservatively in a semi-rigid orthosis for up to 12 weeks. Displaced fractures are reduced followed by a halo-vest instillation. Unstable fractures should be reduced and stabilized by surgery.

Odontoid Fracture

These fracture occur in the young due to flexion from high energy trauma and the elderly due to hyperextension from face flat fall. Patients complain of neck pain and stiffness due to muscle spasm. Radiographical images include plan x-rays of the cervical spine and an MRI to look for rupture of the transverse ligament.

Anderson and D'Alonzo Classification

Type	Description	Stable or Unstable	Treatment
I	Avulsion fracture of the tip of the odontoid from traction of the alar ligament	Stable	Immobilization in rigid collar
II	Fracture at the junction of the odontoid process and the body of the axis	Unstable	Immobilization in Halo-vest; Reduction in cases of displaced fracture
III	Fracture through the body of the axis	Stable	Immobilization in Halo-vest; Reduction in cases of displaced fracture

Sub-axial Fracture

Wedge Compression Fracture

This type of fracture occurs in the vertebral body from excessive flexion. Plain x-rays is enough to make the diagnosis but an MRI or Ct scan may be done to rule out neurological injury. The patient is placed in a comfortable collar for up to 12 weeks.

Tear Drop Fracture

Also known as a burst and compression-flexion fracture is due to axial compression of the cervical spine in divers and footballers. Axial compression results in a burst fracture while a combined axial compression and flexion results in a tear drop fracture where the antero-inferior fragment of the vertebral body is pulled off.

Radiographical images include x-rays of the cervical spine in AP and lateral views with flexion and extension to uncover the vertebral fragment. An MRI or CT scan may be done to look for spinal cord injury. Bed rest and traction followed by a halo vest immobilization is sufficient for treatment of fractures without neurological deficits. In cases where the spinal cord is under threat from a bone fragment an anterior decompression with anterior corpectomy, bone grafting and plate fixation.

Fracture Dislocation

Overriding of the inferior articular facet over the superior facet over the vertebra below results in fracture dislocations where articular masses may be fractures or jumped facets which are pure dislocations may occur. These dislocations may be bilateral and are usually due to flexion-rotation injuries. Spinal cord injury may be possible on resulting rupture of the posterior ligament with instability of the spine. Radiographical images include one of the lateral x rays which shows forward displacement of the vertebra along with an MRI which may be done to rule out the possibility of disc disruption. Immediate management must be preceded with reduction and skull traction with weights according to severity along with IV muscle relaxants. After treatment a collar or a halo vest should be worn for up to 12 weeks.

In unilateral facet dislocation only one apophyseal joint Is dislocated. Spinal cord damage is rare. Radiographical images include lateral and AP x rays of the cervical spine. Management is the same as for bilateral facet dislocation as mentioned above.

Whiplash Injury

Whiplash injuries are usually due to severe motor vehicle accidents of low velocity rear end collision. The head flips backwards and recoils in flexion. Intervertebral discs may be damaged and the anterior longitudinal ligament of the spine and the capsular fibres of the facet joint may be sprained. The patient presents with pain and stiffness which appears in a matter of 1-2 days. Pain radiation to the shoulders along with neurological symptoms and restricted neck movement occurs in this injury. Radiological results usually show straightening out of the normal cervical lordosis but features of long standing intervertebral disk degeneration may occur as the injury progresses.

Management of this injury includes collars and simple pain relief along with minimal range of motion exercises of the neck.

Proposed Grading of Whiplash Associated Injuries

Grade	Clinical Pattern
0	No neck signs or symptoms
1	Neck pain, stiffness and tenderness
2	Neck symptoms and musculoskeletal signs
3	Neck symptoms and neurological signs
4	Neck symptoms and fracture/dislocation

Thoraco Lumbar Fractures

Parameter	Flexion-Compression Injury	Burst Injury	Jack-Knife Injury	Fracture Dislocation Injury
Mechanism	Sever flexion in osteoporotic	Axial	Combined	Segmental

	individuals	compression. Posterior aspect of vertebral body with fragmentation and disc displaced in the spinal cord.	flexion and posterior distraction in lap seat belt injuries	displacement with flexion/compression/ rotation/shear. -Mostly occurs at the thoracolumbar region
Description	1.Patients with minimal wedging and a stable fracture pattern. 2.Those with moderate wedging (loss of 20-40% of anterior vertebral height) and a stable injury 3.Loss of anterior vertebral height >40 % 4.Wedge compression fracture and neurological impairment. 5.Complete paraplegia with no improvement after 48 hours.	Minimal anterior wedging with stable fracture and no neurological damage.	Unstable injury with neurological damage and tensile failure.	Spine is grossly unstable with involvement of all 3 columns and neurological damage associated with the cauda equina. Can present as *fracture-dislocation with paraplegia* or with *a partial neurological deficit*. Can also present as *fracture-dislocation without neurological deficit*.
Treatment	1.Bed rest (2 weeks) 2.Thoracolumbar brace or a body cast with back in extension. 3.Surgical correction and internal fixation. 4.Operative decompression and stabilization through trans-thoracic approach. 5.Conservative management with bed rest followed by gradual mobilization in a brace.	Bed rest followed by mobilization in a thoracolumbar brace or body cast for 12 weeks.	Body cast for 3 months or well-fitting brace.	-Fracture dislocation with paraplegia: Conservative treatment -Fracture-dislocation with a partial neurological deficit: Conservative treatment and/or surgical stabilization and decompression via combined posterior and anterior approach. -Fracture-dislocation without neurological deficit: Surgical stabilization

www.ingramcontent.com/pod-product-compliance
Lightning Source LLC
Chambersburg PA
CBHW041111180526
45172CB00001B/210